Industrial Wastewater Management and Water Reuse Topical Conference

Topical Conference at the 2021 AIChE Spring Meeting and 17th Global Congress on Process Safety

Online
18 – 23 April 2021

ISBN: 978-1-7138-5326-8

Printed from e-media with permission by:

Curran Associates, Inc.
57 Morehouse Lane
Red Hook, NY 12571

proceedings.com

Some format issues inherent in the e-media version may also appear in this print version.

AIChE
120 Wall Street, FL 23
New York, NY 10005-4020

Phone: (800) 242-4363
Fax: (203) 775-5177

www.aiche.org

Additional copies of this publication are available from:

Curran Associates, Inc.
57 Morehouse Lane
Red Hook, NY 12571 USA
Phone: 845-758-0400
Fax: 845-758-2633
Email: curran@proceedings.com
Web: www.proceedings.com

TABLE OF CONTENTS

622759 Water in the Circular Economy Plenary Session

Abstract:

Water in the Circular Economy – Plenary Session

In this inaugural year for the Industrial Wastewater Management and Water Reuse Topical Conference, we have chosen the topic of Water in the Circular Economy for our plenary session. The session will include presentations by three preeminent experts in the field of water and wastewater reuse, followed by a panel discussion featuring all three speakers.

Dr. Nancy Love will present on the topic of Life Cycle Assessments (LCA) for environmental impacts and how they can be used to assess and inform discussions on water systems using a holistic perspective.

Dr. Jaehong Kim will discuss opportunities and challenges for industrial water treatment and reuse from his perspective as a member of the National Alliance for Water Innovation (NAWI) which is a research hub funded by the U.S. Department of Energy (DOE) and led by Lawrence Berkeley National Laboratory. NAWI's focus is on early-stage research on desalination and associated water-treatment technologies to secure affordable and energy-efficient water supplies for the United States from nontraditional water sources. A crucial component of NAWI's five-year research program is an annual road mapping effort designed to engage stakeholders in the water-use and treatment community and inform an annual requests-for-proposal process. This talk will summarize NAWI's roadmap on industrial water treatment and reuse, with special focus on areas of research interest in six industry sectors, food and beverages, primary metals, pulp and paper, oil refineries, chemicals, and data centers and campuses. The talk will also present an example case study on a specific plant with detailed analysis on how modification of water treatment processes can enhance the energy efficiency and lower the levelized cost of water.

Dr. Joseph B. Powell will round out the presentations with a talk on "Energy, Chemicals, and Water." Climate change and circularity are key issues to address in Energy and Chemicals. Doing so will require an unprecedented rate of change in global systems. The varied impacts and demands on water will be examined for some of the key options and scenarios in development to address these grand challenges in sustainability.

We invite you to participate in this timely discussion.

620519 A Self-Cleaning, Durable, High Flux Microfilter

Abstract:

Membrane filtration is typically operated either in a dead-end or cross-flow mode. In these systems the
biggest problem in applications with relatively high suspended solids concentration is low flux, and subsequent fouling or clogging. Because of this, particularly for dead-end systems, the membrane or filter has to be either cleaned via chemical means or replaced often. Other options are based on mechanical self-cleaning filters which do not go to low enough micron ranges, often 2 micron, for reuse
or injection and thus labor and cost intensive alternatives like bag or sand filters are still commonly used. Sofi Filtration Ltd has developed an innovative industrial water filter to address the filtration challenges of small particle size, and high or variable concentration streams. The design criteria for the
new filter was to accomplish a cross-flow filter that is compact and fully automated with a high microfiltration capacity and novel clean-in-place abilities to prevent clogging. The unit requires no internal moving parts to implement the design thus making it robust and economical to use.
The filter is differentiated in its performance from other screens, filters, and bag filters by this design. In
cross flow mode, the feed flow is accelerated within a filter element housing across the exterior of a cylindrical filter element. Over time, as the pressure increases, a self-cleaning cycle is initiated. A high
powered, ultrasonic cleaning element, on the filtrate side, is started to loosen materials which may be causing fouling or clogging the sintered metal screen. After a designated time, while the ultrasound is still on, a back pulse of around 90 psi creates an impact in less than one second to efficiently remove material from the screen. After these 2-steps, the system returns to normal filtration operation. Feedback control ensures appropriate timing of cleaning intervals to maintain filtrate flowrates to manage variable influent solids loadings.
A Sofi Filter unit including a multi-layer metal membrane with nominal pore sizes from 0.5 to 10 microns
has been installed in several applications including saltwater disposal wells for Produced and Flowback
water. Filtrate flowrates up to 2.5 BPM (100 gpm) were measured for a 1 m2 filter element with a nominal 0.5 micron pore sizing. Being of all metal construction without the need for operator intervention due to newly developed smart feedback controls, this system seems poised for more deployments in the upstream and midstream markets.
Furthermore, Sofi Filtration has recently established a technology facility in the Houston area to further
assess the US market applications with Sofi Filter test unit. Test capabilities are initially from 0.5 to 10
micron filtration rating. The facility has full laboratory capabilities for application testing.

619759 Optimization of neutralization process for waste generated from ion exchange regeneration for expansion of a chemical manufacturing facility

Abstract:

Optimization of neutralization of the waste streams generated during regeneration of Ion-Exchange (IX) beds is discussed using a new process such that the acid and base in the waste streams can help self-neutralize, thereby minimizing the use of chemicals required for pH adjustment. The current neutralization setups employed in the industry use excess chemical and energy to neutralize the waste, leading to large quantities of acid and base chemical wastage. A new neutralization process is developed and optimized which is environmentally friendly and cost-effective. For this purpose, various neutralization setups involved in treating IX regeneration waste were studied to reduce energy and chemical usage. From the collected plant data, the treatment requirements of individual waste streams generated during regeneration of IX beds were examined. The currently employed neutralization setup's in the industry are not environmentally sustainable and do not provide proper mixing as a result of which waste large amount of acid and base chemicals. An optimized neutralization setup is developed to lower the operating and capital costs. In the optimized neutralization process, one can achieve adequate turbulence by mixing waste streams generated during regeneration of IX beds. The new developed process is environmentally friendly and economically viable. It is estimated that the new process will result in considerable savings compared to currently used process in the industry. This new treatment process will benefit the chemical industries that require demineralized water for their operation.

620567 MBBR Wastewater Treatment System for Nitrification of Combined Industrial and Municipal Wastewaters under Cold Weather Conditions with Wide Load Fluctuations

Abstract:

Uncontrolled discharge of ammonia in treated effluent from wastewater utilities is harmful to the receiving bodies of water. At high concentrations ammonia is toxic to aquatic life. It also adversely impacts the presence of dissolved oxygen which is critical to the health of water bodies. Additionally, discharge of nitrogen in treated wastewater in all forms, including ammonia, promotes undesirable growth of aquatic plants e.g., hyacinth, algae, etc., also called eutrophication, which renders water unsuitable for use and recreational purposes. Environmental regulators throughout North America and globally are increasingly calling for effluent nitrogen control, which requires nitrification as the first step. Nitrification involves biological oxidation of ammonium ions in wastewater. It is kinetically a very slow process and sensitive to environmental conditions, especially wastewater temperature, pH, presence of heavy metals and other toxicants. Many treatment plants, especially in rural US and Canada are lagoon based that can remove only organic pollutants - BOD (biological oxygen demand). In winter, these lagoons are exposed to very low ambient temperatures. Nitrification is difficult to sustain under cold temperatures <5°C. Moving Bed Biofilm Reactors (MBBR) have been shown to effectively conduct nitrification under sustained low temperature conditions while consistently achieving low ammonia limits. Other benefit of the MBBR process include – (a) automatic growth and retraction of the biofilm under varying flows and loads without having the operator to adjust, monitor, or control parameters such as recycled activated sludge (RAS) and mixed-liquor suspended solids (MLSS), (b) requirement of much smaller reactor volume and footprint area, and (c) ability to quickly recover from hydraulic surges, and toxic and shock loads.

Batesville, AR, has a population of around 10,000 but its wastewater treatment plant treats 60,000 population equivalent flows due to many industrial facilities in the area including poultry and dairy industries. Prior to 2015, lagoons were used to treat the wastewater to remove only BOD. Lagoons were also used to settle the solids. Due to substantial growth in industrial discharges to the facility, the plant was stretched to the limit to remove BOD, especially in winter, when the water temperature in the lagoons reach as low as 4° C.

In 2015, the plant's permit was revised to remove ammonia-nitrogen with a year-around limit. Due to the very low water temperatures in winter expansion of the lagoons was not considered a viable option. MBBR was selected as the technology to meet the new regulatory requirements for BOD and ammonia-nitrogen control due to its ability to perform under very low temperatures and handle varying and toxic loads from industrial contributors.

This presentation will consist of an analysis of many years of operating data demonstrating how the MBBR process helped the treatment plant achieve its goals of producing stable effluent quality consistently over the year even under low temperatures and varying loads. Operating performance will be compared with kinetic models based on the principle of MBBR technology.

NOTES

NOTES

NOTES

NOTES

NOTES

NOTES

AUTHOR INDEX